YOUR KNOWLEDGE HAS VALUE

Renard Teipelke

San Diego's Bioscience Cluster and Berlin's Club Scene

The Scope of Urban Policies in a Florida-esque Age of Creative Cities

GRIN Verlag

Bibliografische Information der Deutschen Nationalbibliothek:

Die Deutsche Bibliothek verzeichnet diese Publikation in der Deutschen National-
bibliografie; detaillierte bibliografische Daten sind im Internet über http://dnb.d-
nb.de/ abrufbar.

Imprint:

Copyright © 2012 GRIN Verlag GmbH
Druck und Bindung: Books on Demand GmbH, Norderstedt Germany
ISBN: 978-3-656-46306-1

This book at GRIN:

http://www.grin.com/en/e-book/230180/san-diego-s-bioscience-cluster-and-berlin-
s-club-scene

GRIN - Your knowledge has value

Der GRIN Verlag publiziert seit 1998 wissenschaftliche Arbeiten von Studenten, Hochschullehrern und anderen Akademikern als eBook und gedrucktes Buch. Die Verlagswebsite www.grin.com ist die ideale Plattform zur Veröffentlichung von Hausarbeiten, Abschlussarbeiten, wissenschaftlichen Aufsätzen, Dissertationen und Fachbüchern.

Visit us on the internet:

http://www.grin.com/

http://www.facebook.com/grincom

http://www.twitter.com/grin_com

Renard Teipelke, 2012

Veranstaltung:	Seminar Cultural Geographies of Economy
Abgabetermin:	16.03.2012
Autor:	Renard Teipelke
Studiengang:	M.A. Geografien der Globalisierung

San Diego's Bioscience Cluster and Berlin's Club Scene:
The Scope of Urban Policies in a Florida-esque Age of Creative Cities

Table of Content

Abstract

This paper discusses the relation of Richard Florida's notion of creative cities, cluster economics, and urban policies towards creative industries. Two empirical cases, San Diego's bioscience cluster and Berlin's club scene, are examined in order to reconstruct their success, analyze corresponding factors, identify challenges and problems, and discuss recent developments. Conclusions will be drawn on what city governments' policies can or cannot as well as should not do to support creative industries. It will be argued that the scope of urban policies even in a Florida-esque age of creative cities is still well-related to ('traditional') cluster policies but therefore also limited by scale and dependent on private enterprises and cluster actors that have an active role in attracting other creative people.

1. Introductory Remarks

1.1. Setting the Stage

This year we are celebrating the 10th anniversary of Richard Florida's groundbreaking book "The Rise of the Creative Class" (2002). For some, it will be a real celebration, for others it will be a day of mourning for one decade of ill-suited, big-government or neo-liberal urban policies – depending on the political standpoint. Even though Florida's ideas were not completely new, he changed the thinking and acting of many city officials worldwide. It has been his style of presenting and promoting the concept (rather than a 'theory') of the creative class. And civic leaders have since then been eager to hear what role creative workers can and will play for the socioeconomic vitality of cities and regions.

Briefly summarizing one of Florida's main points that has not been totally refuted, I would like to underscore that "geography is not dead" (Florida 2002: 4), and thus places have to attract high-skilled creative people that will move to cities which can offer a high quality of life. Even though Florida wrote that "there is no one-size-fits-all strategy" (2002: xxiii), he has been giving various recommendations to city officials on what they have to do in order to attract the talent that will ensure their cities' growth (or survival) in the future. Florida explains the importance of culture and its scenes (music, film, art, outdoor recreation etc.) – traditional amenities such as shopping malls or sports stadiums do not attract the creative class any longer (2002: 9, 183). Instead of these physical attractions, talented people seek "high-quality experiences, an openness to diversity of all kinds, and, above all else, the opportunity to validate their identities as creative people" (Florida 2002: 9). What already becomes apparent is that even these recommendations lack specific guidelines on 'how' city officials can eventually achieve the newly outlined goals.

This might be due to the specificity of each city's case, but the following discussion of Florida's ample critics will underscore how 'good old urban policies' have been disregarded in a heated debate characterized by many stereotypes of and accusations against social-liberals versus social-conservatives as well as government-advocates versus neo-liberal hardliners (cf. Peck 2005: 740-741).

1.2. Refuting Florida, Praising Governance, Sidelining Cluster Economics

The format of such a paper is far too limited to give a complete overview of the responses to and critiques against Florida that have filled hundreds of journals since 2002. Therefore, I will not deal with the methodological objections against Florida's use of statistics and rankings (cf. Malanga 2004, or Peck 2005 criticizing Malanga's use of statistics to refute Florida: 755), because they will not be part of my argumentation. Furthermore, I will also not deal with

Florida's alleged political motives and possible negative repercussions from his policy recommendations that he did or did not neglect (cf. Peck 2005: 746).

What seems most relevant to me are three aspects: the logic behind Florida's argumentation, the advocacy for (self-) governance in the creative economy, and the sidelining of cluster economics in much of the creative city research. Starting with the first aspect, Peck (2005: 755), amongst others, correctly underscores how the circular fashion in Florida's argumentation does not shed light on the (direction of the) causation between economic growth and cultural innovation. Another corresponding critique by authors such as Pratt (2008) is that the understanding of cultural activities as a somehow disregarded business in the past has led cities to co-opt culture into (neo-liberal) urban strategies for economic development (catchphrase: commodification of arts).

The second aspect deals with authors that accept and appreciate the role of cultural industries in a post-Fordist era. Amongst others, Lange et al. (2008, 2009b) have spent a great amount of work understanding and analyzing features of creative (cultural) industries and explaining what these characteristics mean for traditional urban politics: "informal alliances between private and public stakeholders, self-organized networks to promote new products in new markets and context-oriented forms such as branding of places, represent new forms of managing the urban" (Lange et al. 2008: 535). Since creative industries contrast typical hierarchical power and representation structures, the solution for 'creative cities' to support these industries lies in "context governance" in which a supportive environment is created (Lange et al. 2009b: 26). Self-governance of creative industries such as in networks or co-governance such as in public-private partnerships (Lange et al. 2009b: 16-18) could be understood as a practical response to Florida's request for an enriching urban climate that encourages creative people to realize their ideas.

The third aspect is directly related to the second one, because in the academic discourse about new forms of urban governance, traditional ideas of urban government have become less salient in the debate. Critics like Peck (2005: 767) are arguing that the general assumptions derived from (neo-) liberal economic theories are already false and thus first have to be radically changed before we can discuss sustainable approaches to high-quality urban life. However, this revolution might still take some time and scholars would exclude themselves from discussions about what city governments can do right now in order to achieve various goals. If we accept the premise that cities are in competition with each other (not for the sake of competition itself, but because of the disposition to differentiate themselves from each other – which does not rule out inter- or intra-regional cooperation), cities have to appeal to creative industries and people that will form the foundation of the socioeconomic well-being of urban regions (Scott 2006a: 2, 5, 13).

Therefore, I cannot explain why the extensive stock of cluster research has not often found its way into the discussion about creative cities with regard to urban policies. If we understand governance as governing, coordinating, or managing (the process of) politics, often with a specific focus on actors, institutions, and networks as well as their interrelations (Benz and Dose 2010: 25-27), the overwhelming majority of literature deals with governance aspects. Thereby, it has not fulfilled the task of bringing new understandings of governance and findings from research on urban policies towards clusters together. At the end of my paper, I will again refer to these perspectives and related tasks for research, but for the moment now, I would like to stress the following: Without any doubt, governance aspects have and will increasingly become more relevant. Nevertheless, 'traditional' government is still there and I think it is worthwhile to critically assess the scope of policy options with regard to creative industries. At this juncture, it might also be worthwhile to look at ('traditional') cluster policies, because creative industries of a particular (economic) strength/size are often very similar to clusters with their spatial proximity, business network structure, information exchange systems, internal relationships, particular infrastructures, and the spatiality of knowledge creation (cf. Evans 2009; Bathelt, Malmberg, and Maskell 2004).

1.3. Paper Outline

Having laid out the research field as I perceive it, the general questions and related hypotheses for this paper are very clear and simple: I will investigate two empirical cases, the bioscience cluster in San Diego (United States) and the club scene in Berlin (Germany), in order to reconstruct how they became essential creative 'sectors' of their city. I will analyze main success factors, deal with challenges and problems, and discuss recent developments with regard to these creative industries and their corresponding city governments. Then I will draw conclusions from the examples by assessing what city governments can or cannot as well as should not do with respect to urban policy options when trying to support creative industries. My main argument is that the scope of urban policies even in a Florida-esque age of creative cities is still well-related to ('traditional') cluster policies but therefore also limited by scale and dependent on private enterprises and cluster actors that have an active role in attracting other creative people.

With regard to the two empirical cases, I understand 'creative industries' as being creative insofar as they are fundamentally dependent on creative individuals that are constantly developing innovative ideas in science or arts. People working in these creative industries need an enriching and supportive environment, which means that place-related factors are of high relevance (cf. Florida 2002: 44-48).

While San Diego has more to offer than only its bioscience cluster (defense and transportation, cybersecurity and robotics, energy and cleantech, information technologies and communications; CONNECT 2012b: 7-11), Berlin's creative industries go far beyond the club scene (fashion design, film, software and games, design and architecture, advertising, and performing arts; SenWTF 2008: 24), I decided to focus on biotech and the club scene, because I have lived in both cities and experienced as well as studied the biotech cluster and the club scene in the everyday life (University of California in San Diego campus in Torrey Pines Mesa; podium discussions and nightlife) as well as in my professional occupation (field study on sustainable biotech facilities in San Diego; work for a participatory neighborhood management and branding firm in Berlin).

Even though San Diego's bioscience cluster and Berlin's club scene seem to not have much in common, the following analysis will show how, for example, Florida's 3 T's (talent, technology, and tolerance; 2002: 10-13) can be found in both examples. That the two cases are based on different forms of these 3 T's is without any doubt. Nevertheless, both cities are regularly named as creative places, brand themselves as bioscience or club/music cities respectively, and are trying to adjust their policies in order to support their corresponding creative industries (cf. for instance the official websites of both cities).

2. Empirical Cases

2.1. San Diego's Bioscience Cluster

The Story

San Diego's bioscience cluster is one of the best in the United States (after Boston and San Francisco) and home to multi-billion-dollar transnational corporations as well as ample start-up companies that together make up a majority of the region's 44,000 employees and 6,400 patents (CONNECT 2012b: 7-9). Concerning the term 'bioscience', it can be thought of as a mixture between biotechnology and life science thus resembling the concentration of biomedical research and production in San Diego.

Walcott (2002) introduces her study of the bioscience cluster in San Diego by describing it as an "innovative environment and the matching of place characteristics with a specific economic activity" (99) and explaining how the "formation of synergistic connections" promoted "political, economic, and social networks of entrepreneurial individuals at the metropolitan scale" (99). Even though I will not go into the details of cluster research with regard to innovation transfer etc., I will briefly describe the rise of San Diego's bioscience cluster before identifying major factors of its success.

Started with the founding of the University of California in San Diego (UCSD) (and affiliated as well as) adjacent research institutes (Scripps Institute of Oceanography, Scripps Research Institute, Salk Institute for Biological Studies, Sanford-Burnham Medical Research

Institute), the bioscience cluster really took off after the success of Linkabit (1968; the multi-billion-dollar telecommunication corporation Qualcomm became its most important spin-off) and the anti-bodies producer Hybritech (1978) (more on the history, cf. Global CONNECT 2010, Walcott 2002). Both companies were founded by UCSD professors, which already hints to the strong interrelation between research and risk business in San Diego, or the Torrey Pines Mesa community in particular (Bennett 2008). Having been far away from the bioscience centers in Boston and San Francisco as well as the political centers in Sacramento (state level) and Washington, D.C. (federal level), San Diego nevertheless recorded a significant increase in new companies, jobs, and investments in the 1990s (Walcott 2002: 102). This is often accredited to the region's open communication culture, knowledge economy, and "the promise to build something from nothing" (Global CONNECT 2010: 7) in a place where newcomers are always welcome and integrated. But we have to look closer into the details in order to identify what the actual success factors have been.

Factors

San Diego's bioscience cluster could not exist without the constant knowledge and talent 'production' at the "outstanding research university" UCSD (Walcott 2002: 99) and its adjacent research institutes. Early on, they developed an environment in which competition and cooperation as well as an interdisciplinary and entrepreneurial approach have been highly appreciated and advocated (Global CONNECT 2010: 4). This human and physical capital (creative talents and cutting edge laboratories) have been met with billions of dollars from both the private sector (venture capital) and the state and federal budget (R&D grants, particularly from the National Institutes of Health) (Global CONNECT 2010: 4; Walcott 2002: 102). One has to underscore the Bayh-Dole University and Small Business Patent Procedures Act of 1980 – an intellectual property legislation that fundamentally facilitated the partnership between research institutes or their scientists with private enterprises (Walcott 2002: 105).

The bioscience cluster has also profited from San Diego's relative remoteness, since this has necessitated and strengthened human networks and a collaborative community that resulted in a region-based milieu (Global CONNECT 2010: 8). Information has been exchanged in this network intensively and newcomers have found plenty of new 'playing fields', since the region, even though specialized in life science, still has had a diverse range of sectors to offer for new innovations (Global CONNECT 2010: 4). Thereby, the cluster in San Diego relied time and again on private-led initiatives to improve the economic health of the life science sector. CONNECT as the non-profit life science organization and the San Diego Biocommerce Association (BIOCOM) are only the two most prominent examples to share information, support research and start-up enterprises, connect them to capital, offer training, represent the industry, and advocate cluster-related policies on every policy level (CONNECT 2012a; BIOCOM 2006, 2010).

Finally, there are also place-specific factors for San Diego's bioscience cluster, as for example the spatial proximity in Torrey Pines Mesa that has only been possible, because the city set aside plenty of (though not enough) industrially zoned land (Walcott 2002: 110). The cluster has also been dependent on a secured supply of large amount of water as well as a high-quality energy infrastructure that was originally realized by heavy investments of the State of California (Bennett 2008).

Challenges

For an extensive cluster analysis, one would have to describe the challenges San Diego's bioscience cluster has been facing with much more details. For the purpose of this paper, it shall be sufficient to note the following challenges that are directly related to the cluster's success factors: The bioscience cluster sees its talent reservoir threatened, because the fiscal crisis of the State of California has dramatically worsened in the past years of the financial/economic crisis (Global CONNECT 2010: 13). A de-investment in public universities (like UCSD) and K-12 education could severely disadvantage the region in its national as well as global competition – even more with regard to San Diego's challenge to integrate immigrants (Walcott 2002: 108).

Another problem regards the region's transportation infrastructure, particularly the linkages: Traffic gridlock and a rather old and small airport do not support the interconnection of San Diego with other clusters elsewhere (Global CONNECT 2010: 7; Walcott 2002: 108). The shortage of water and affordable (industrially zoned) land will be an even bigger challenge, since those two factors concern natural/geographic aspects of the region and San Diego in particular and have inhibited the mature-stage development of large-scale manufacturing and distribution facilities (BIOCOM 2006, 2010). The shortage of developable land has also fiercely decreased affordable housing in San Diego (City of San Diego 2008: SF-27). With regard to the state level, CONNECT, amongst others, criticizes life science regulations and tax rates (Global CONNECT 2010: 13).

Recent Developments

The case of San Diego is of great importance in 2012, because the city will hold mayoral elections this summer. Candidates already invented slogans that aim at reigniting the 40-year-old city proclamation "America's Finest City". The new slogans, such as "The World's Most Innovative City" (Fletcher 2012: 3) or "Perfect Climate for Business" (DeMaio 2012: 85), underscore how San Diego and its creative industries are thought of as one entity in branding the city. More companies (or people) shall be attracted by a high quality of life in an environmentally prime location. Easier and faster permit processions, fee waivers, an improved public transit system, job coordinators for specific industry sectors, the enlargement of the convention center, as well as an intensified advertising of the city and its economy through fairs, exhibitions, and 'diplomatic missions' are all on the agenda of the candidates (cf. Voice of San Diego 2012; Fletcher 2012; DeMaio 2012).

San Diego's city officials seem to have also read some of Florida's recommendations or simply come up with the following ideas on their own: As a minority majority city, San Diego increasingly praises its American-Mexican history and culture (City of San Diego 2008: iii). The City of San Diego Commission for Arts and Culture's Organizational Support Program, which is funded by the transit occupancy tax, supports local culture as does the "Public Workers Projects 2% for Art" and the "Private Development 1% for Art" (City of San Diego Commission for Arts and Culture 2011; City of San Diego 2008: UD-30-33). With a "city of villages" strategy (City of San Diego 2008: SF-3), San Diego has developed neighborhood-specific mixed-use centers and follows plans to further extend the public transit system as well as its pedestrian-friendly urban design.

The often appraised environmental features of San Diego (parks, beaches, canyons etc.) play a major role in the city's general plan (City of San Diego 2008: SF-6, 20-21, 23-24, CE-4-7, RE-19-36), in which sustainability goals are indicated with respect to, for example, the reuse of reclaimed water, conservation measures, or the implementation of a green network of open spaces that shall further encourage recreation activities. From the previous explanations, several conclusions will be drawn after the second empirical case is presented.

2.2. Berlin's Club Scene

The Story

Taking Berlin's club scene as an example for creative industries is a quite daunting task, because it is very difficult – if we compare it, for instance, with San Diego's bioscience cluster – to find exact figures to describe the nature and role of the club scene in and for the city. While most reports subsume parts of the club scene under "music industry" (SenWTF 2012; House of Research 2011), two publications by the Senate Department for Economy, Technology, and Women Issues, a study on the economic potential of Berlin's club and event sector (SenWTF 2007) as well as the creative industries report (SenWTF 2008) have been more specific. The latter report, for instance, lists 120 clubs, 236 music publishers, 94 recording studios, 65 concert and event management enterprises, and 21 concert and event stages (SenWTF 2008: 105). The music industry is supposed to make revenues of around one billion Euros per year through live events, festivals, and the clubs (Stollowsky 2011).

Taking into account that the scene is still characterized by a significant minority of informal arrangements, a picture of the exact nature of Berlin's club scene will stay relatively fuzzy.

Berlin has a leading role with regard to various creative industries and the importance and impact of its club scene summarizes Rapp (2009) as follows:

> This new Berlin that attracts thousands of clubbing tourists every weekend is the party capital of the Western world. It is a city with low rents and liberal public authorities. The reality principle of other cities is deferred in favor of an eclectic lust principle. Nobody has to actually work here, if not in some art or music projects. New clubs are opened incessantly and, in fact, people are clubbing all the time. This is really it. (own translation; 33-34)

The now world-famous electronic music club scene actually first emerged during the 'limitless time' after the fall of the Berlin Wall in 1989 in the area between Potsdamer Platz and Friedrichstrasse (with clubs such as WMF, Tresor, E-Werk), before many new clubs – formal, but most often informal – were opened and a location-change happened with clubs moving from the rather central border strip area to the border strip area at the Spree riverfront in the Eastern part of the city (Rapp 2009: 31). Major clubs that have been most prominent since then are Berghain, Weekend, Tresor, Bar 25, Watergate (amongst others…).

This evolution of a multifaceted subculture was accompanied by electronic music festivals like the Love Parade, the foundation of various new labels, producers and their old labels moving to Berlin, the evolution of a fruitful relationship between Berlin's electronic music scene and Detroit's, as well as the establishment or advancement of electronic music subgenres (particularly, house, techno, and minimal) (more on the history, cf. Rapp 2009: 27-70).

In the past years, the club scene entered a new stage when more and more tourists were coming to the city solely to party for a weekend (or a couple of days), thus making Berlin the Mecca of techno culture (at least) in Europe (Rapp 2009: 82). The city also started to brand itself with this diverse scene and the economic impact became increasingly apparent with circa 20 million tourist overnight stays in 2010 (50% of the tourists younger than 30 years; Stollowsky 2011) and corresponding revenues for ho(s)tels, transportation enterprises, shopping malls, bars and restaurants, kiosks and fast food stands, as well as clubs (Hollersen and Kurbjuweit 2011; SenWTF 2007).

Factors

The reasons for Berlin's establishment of the club scene are a mix of historical aspects and 'hard' and 'soft' factors. The fall of the Berlin Wall opened up ample developable land in the former border strip area (Rapp 2009: 31). Club owners chose those places where either a development plan had not yet been designed or no private (or even public) body wanted to invest. In the course of the years, creative people of the club scene found empty plots or vacant (derelict) buildings they could (often temporarily) use for low rental prices that have really been a crucial locational factor especially for producers and DJs (Rapp 2009: 31-34). These places were redesigned/reconstructed in order to have places for production and performances close by (SenWTF 2008: 110, 114-115). The club scene was able to create something what Florida has termed "audio identity" of a place (2002: 228).

This was the foundation of a community, in which music producers have benefited from spatial proximity to other cultural industries in the same neighborhoods as well as from the related dense cultural programs (House of Research 2011: 23). Different subcultures have built networks in which information could be shared easily and new ideas developed. 'Art' producers have found an open-minded audience (House of Research 2011: 23). Alternative lifestyles have been (indirectly) supported by the city that came to be known as liberal and gay-friendly – or in the words of Florida: Berlin's image has been "openness to diversity of all kinds" (2002: 9). Corresponding politics, such as a liberal drug policy (jof/dino 2010), was complemented by cheap flight airlines (like AirBerlin, Easyjet, or before HLX) that have facilitated international exchange and made weekend traveling to Berlin affordable (Rapp 2009: 79-80).

Challenges

Problems affiliated with the club scene are manifold. However, as I did not undertake a complete cluster analysis in the case of San Diego's bioscience cluster, I will also not discuss the benefit or detriment of the club scene and its subculture, but will concentrate on challenges that the scene is facing (particularly club owners and producers).

A major characteristic of the club scene, its temporary nature, is becoming more and more a challenge: Rental agreements are short and rental prices are rising (Zha 2011), and nevertheless, owners have to make significant investments in their clubs in order to meet standards – set by official regulations (fire, noise, security laws), expected by the audience, or defined by the owners' and DJs' own ideas of qualitative clubs (Rapp 2009: 33, 48; House of Research 2011: 28; SenWTF 2007: 17).

At the end of last year, it happened again that some prominent clubs had to close down for rental or financial reasons (catchphrase "Clubsterben" – 'the dying of clubs'; Rapp 2009: 51), which sparked yet again a loud outcry in the club scene against politicians and city officials accused for not really supporting the club scene or understanding the related economic value as well as the value chain in the alternative music scene (Stollowsky 2011; House of Research 2011: 28; SenWTF 2007: 3-4, 16). This is interrelated with a representation problem of the club scene that is not organized in a traditional hierarchical power structure and thus has less leverage in (city) politics (Bürkner 2009: 253-254). This became also apparent recently when tax authorities questioned the reduced value-added tax for club entrance fees, implying that a DJ's work cannot be treated like a concert or other tax-reduced art performances (Mösken 2011; Stollowsky 2011).

The protests against the alleged neo-liberal riverfront development project "MediaSpree" is another example of club scene stakeholders (together with other creative people) criticizing the city government's alleged neglect of gentrification tendencies in Berlin, particularly in neighborhoods where alternative scenes currently have their home base (Mösken 2011; SenWTF 2007: 4; Rapp 2009: 37-70). The privatization of public space will furthermore pose a tremendous challenge to the club scene (Zha 2011).

With the increase of weekend travelers ("Easyjetraver"; Rapp 2009: 78), two other problems have become more difficult to solve: Clubs are increasingly struggling with noise protection laws that request fortified noise protection installations in the clubs, while more tourists also mean more noise in front of the clubs (also due to non-smoking-laws) (Stollowsky 2011). I will further discuss the corresponding challenge that "partytown is sucking the authenticity out of the city" (Hollersen and Kurbjuweit 2011) again in chapter 3.3.

Recent Developments

Even though the accusations of stakeholders in the club scene are relatively harsh against city officials, the city government has already introduced programs to support Berlin's creative industries. The main program has been "Projekt Zukunft" ('Project Future') through which economic structural changes should be implemented in order to facilitate job creation in the most promising (creative) sectors (SenWTF 2012). In 2004, two elements of the project started that directly targeted creative industries and aimed at changes in the bureaucratic processes to better meet the needs of creative entrepreneurs and foster a better connection as well as cooperation between creative industries. The project has also included investment programs and discussed leaner credit lending for start-ups as well as support for trade fair presentations (SenWTF 2012: 8-11). The online platform "Creative City Berlin" was launched and support for temporary uses of vacant buildings has been promised (and partly realized) (SenWTF 2012: 10).

On the side of club scene stakeholders, the Clubcomission Berlin was founded in 2000 as an association of club owners as well as party and event managers in order to operate as an interest group/mouthpiece of the club scene (though several of the most important clubs are not members). Besides, the Clubcommission Berlin offers training, organizes discussion panels, gives legal advice, and takes care of public relations (Rapp 2009: 51-54). Various stakeholders in the club scene have requested that the city government ensures that particular spaces are kept open to creative uses for affordable rental prices, even if it has to purchase them from private bodies (Stollowsky 2011). Furthermore, constructors of new buildings shall be made responsible for fortified noise insulation when building right next to an existing club (Stollowsky 2011). A leaner interpretation of laws and regulations on the part of trade, building and planning, environmental protection, as well as regulatory and health government agencies has also been requested (SenWTF 2007: 17).

As a reaction to the closing of clubs, the existing differentiation between 'high-brow' culture (such as performing arts in opera and theaters), which receives and is highly dependent on public funding, and 'low-brow' culture (such as the club scene), which is not eligible for public funding and often financially sound, has been questioned (Walter 2011). However, ideas for an emergency fund for clubs which run danger of being closed (Groove 2012) and arguments against any peril of losing its independence are currently discussed in the club scene (Stollowsky 2011).

3. The Scope of Urban Policies

In this chapter, I will bring together the two empirical cases by drawing conclusions on what city governments' policies can, cannot or should not do in supporting creative industries.

3.1. Beyond the Scope of Urban Policies

To start with a rather odd notion, I would like to point out how the creative industries in San Diego and Berlin are examples for success factors that are far beyond the control of any actor: In its report on San Diego's bioscience cluster, Global CONNECT mentions the region's excellent and warm weather as one factor that attracts scientists to San Diego (2010: 8). Anyone who has been to San Diego can probably understand this argumentation. And in the case of Berlin, it was the fall of the Berlin Wall that dramatically changed the everyday life in the city and opened up a dynamic that spurred and is still driving Berlin's club scene (cf. Rapp 2009). For weather as well as (inter-) national historical/political changes, there are ways to indirectly influence them; however, these two factors, amongst others, can hardly be formed or governed through urban policies.

Considering the others factors, it becomes more a discussion of political scale, especially in the case of San Diego's bioscience cluster. I have outlined how its success was based on state investments in the infrastructure, as well as the related provision of ample water and energy (also by the state or, at least, the Metropolitan Water District of Southern California). Furthermore, the cluster has depended on state and federal funds for research and development. Another aspect is the research university and its affiliated institutes – as it is the case in the United States (as well as in Germany), the de-/ investment in universities is decided on the state level. One aspect that is seen as a challenge to the bioscience cluster as well as the club scene are laws and regulations, as well as taxes (Global CONNECT 2010: 13; SenWTF 2007: 17) – again policies predominantly decided on the state or federal level, even though in Berlin as a "Stadtstaat" ('city state'), we have a special case in favor of state policies tailored for the political necessities of a city. Nevertheless, what becomes clear is that, in general, many parameters that can support (creative) economies are not managed in the city government.

If we take a closer look at other factors that have been decisive, we can also see that the second pillar of funding for the bioscience cluster is venture capital, thus concerning the private sector. Discussions about rising rents on (industrially zoned) land partly regard privately-owned land (in San Diego and Berlin). Furthermore, an encouraging environment to test out new ideas – in both empirical cases – has been evolving from actors in the corresponding networks (Global CONNECT 2010: 8; Walcott 2002: 111; Rapp 2009: 200-203; Lange et al. 2008: 539-542). Thus, people in the creative industries, not urban policies, create a culture of taking risks and trying out innovations in practice.

With regard to the specific aspect of noise protection laws in Berlin, I would also like to point out that it is not the regulation per se that poses a challenge to club owners but local residents that are advocating against 'loud clubs' in their neighborhood and are trying to disturb clubs by various 'guerilla actions', thus not only acting against the alternative scene but actually hurting the club owners' business (Popblog 2011).

3.2. What City Governments Can Do

Even though various authors correctly conclude that the scope of (city) policies is rather limited (Thierstein, Förster, and Lüthi 2009: 75-76), one can identify action fields from the two empirical cases: City governments can support creative industries, particularly their formation, by 'good old' cluster policies (Evans 2009: 128). These include reducing certain taxes, offering fee waivers, or sharing the costs of new infrastructure provision and site utilization (Scott 2006a: 12; Fletcher 2012: 18). City governments also have a say over permission processing and the interpretation of regulations – all aspects that account for bioscience clusters (for instance support in writing environmental impact reports) as well as the club scene (for instance a liberal drug policy). With respect to those cultural industries that are dependent on public support or funding, one can find various standpoints either proposing certain protection mechanisms that can balance profit-paradigms (Bürkner 2009: 257) or questioning how solely giving out money can ensure sustainable long-term changes (Lange et al. 2009a: 330).

A highly relevant policy field concerns urban development with respect to zoning and public/private land. It might be the field a city government can best exert influence on. In the case of San Diego, it has been the donation of land formerly owned by the city to UCSD as well as ample industrially zoned land close to the campus (Walcott 2002: 110; Global CONNECT: 12). In Berlin, it has been (insufficient) riverfront regulations and rental laws that have aimed at preventing (an even more severe) replacement (gentrification?) of alternative scenes – but failed, depending on the viewpoints (Zha 2011; Mösken 2011). Cities have a huge influence on how they keep public places open to (unintended) uses (Kunzmann 2009: 42-43; Stollowsky 2011). Until now, residential and commercial uses have often not been allowed through zoning; however, it hints to the potential of adjusting zoning regulations to the needs of creative (in this case: cultural) industries (Markusen 2006: 17).

Concerning other policy fields, the scope of city governments varies. Infrastructure with regard to transportation, especially public transit systems, are often planned by city or regional governments, though are often difficult to implement due to their tremendous costs. A similar issue is education that in many cases is regulated by the state, but can be adjusted by city governments to local needs. City governments can also implement supplementary education or specialized training programs in order to provide a talented stock of young people and to ensure that citizens decide to stay/move to the city because of its high-quality education system (Dillon 2012; Scott 2006b: 15).

Thinking about Florida's recommendations towards "quality of place" (2002: 232), city governments have certain influence on the quality of life in their urban regions. Let it be neighborhood beautification initiatives, park conservation programs, plans for vast green open spaces, or the support of local culture through culture commissions, as we have seen in San Diego and Berlin. This can be accompanied by tax revenues that go directly into culture funds (Markusen 2006: 15, 29).

I will not go into the details of what some cities have termed 'enabling policies' and explained with network building, since these aspects deal again with governance aspects. Nevertheless, the network building becomes relevant for my analysis with regard to what Bathelt, Malmberg, and Maskell described as "global pipelines" (2004). While clusters (of creative industries) often will take care of integral information exchange networks, an interactive knowledge creation and learning for themselves ("local buzz"; Bathelt, Malmberg, and Maskell 2004: 38, 41-42, 48), city governments can support the building of "global pipelines", which are the outside connections of clusters to other companies or clusters in their sector (Bathelt, Malmberg, and Maskell 2004: 48). This can be done by, for example, city governments' promotion of trade fairs and exhibitions, the development of regional trademarks, or 'diplomatic missions' to other places on behalf of the city and its creative industries (Scott 2006b: 15-16).

One highly relevant basis on which the execution of urban policies depends is a city government's bureaucracy. Therefore, some authors have called for the qualification of city government officials that have to understand the needs of creative entrepreneurs (Lange et al. 2009a: 330; Markusen 2006: 16; SenWTF 2007: 17). As we can see in the example of the bioscience cluster and the club scene, these needs can be very different and sector-specific. I strongly agree with Berridge (2006), who demands from self-branded creative cities that they start to be creative in their own governments first by finding new ways to deal with challenges such as those which I outlined earlier.

3.3. Where City Governments Should Be Cautious

So far, I have shown that similar conclusions can be drawn from both empirical cases, San Diego's bioscience cluster as well as Berlin's club scene. For some sectors, such as bioscience, it is encouraging and helpful to be put on the main stage in city governments' policies and image campaigns. Other cases, such as the club scene, are better off in their niche and are at risk to lose their alternativeness when standing under the spotlight of cities' branding campaigns (Rapp 2009: 53). Therefore, the Berlin case in contrast to the San Diego example has a feature that is really specific and needs to be highlighted: it is the discussion about what city governments should better not do with regard to creative industries – in our case: the club scene with its alternative music and subculture.

From my research, I would conclude that Florida has not (completely) understood creative (cultural) industries when writing: "The 'counterculture' was – and is – just popular culture, and popular culture is a ticket to sell things and make money." (2002: 200). Without any doubt, club owners see their clubs not only as part of their cultural expression, but also as an investment and business. Nevertheless, there is still a difference between popular culture/music and alternative culture/music. For example, the Berlin club scene does not have a celebrity/star system like the mainstream culture that is heavily focused on commercialization (Rapp 2009: 9-18). The electronic music scene in Berlin intentionally plays with its multifaceted nature of hundreds of clubs, even more DJs, and constantly produced innovations. Living the alternative concepts of life is an essential part of the scene (as exemplified by the story of Bar 25; cf. Rapp 2009: 162-182).

I am describing these characteristics at length, because the increasing rush of weekend travelers ("Easyjetravers"; Rapp 2009: 78) into Berlin has already started to threaten the uniqueness and authenticity of Berlin's club scene, since neighborhoods are experiencing a touristification. The balance between local/regional visitors and tourists in some clubs is heavily changing to the latter group, and residents have started to organize against the tourist inundation as exemplified by Berlin's Green Party roundtable (in famous Kreuzberg's

Wrangelkiez) themed "Help, the Tourists are Coming" (Hollersen and Kurbjuweit 2011; Teipelke 2011; Markusen 2006: 27; SenWTF: 15).

Back to the original question what city governments should better not do: It might have been a major mistake of the city government to brand itself so intensively with Berlin's club scene, before understanding that alternative cultures in a creative niche industry can lose their attractiveness to its part of the creative class (music producers amongst others) when it is experiencing a mainstreaminization. Some creative industries want and need to be not only creative but different – in this case, it can be a supportive 'policy' of the city government to be rather passive and temporarily 'ignore' the scene (Lange et al. 2009a: 330).

4. Concluding Remarks

In this paper, I have analyzed two empirical cases, San Diego's bioscience cluster and Berlin's club scene, in order to draw conclusions on the scope of urban policies. As the previous chapter has shown, this scope depends heavily on the policy field and its corresponding scale. Furthermore, there are factors that are influenced not by public but private actors. Peck (2005: 765) correctly criticizes how cities in Florida's argumentation are made responsible for attracting creative people (and industries) when their influence on many policy issues is actually very limited. Nevertheless, both empirical cases could show that city governments do not completely stand helpless in face of the challenge to develop a sustainable socioeconomic basis for their future.

After having interviewed stakeholders from Europe, North America, South-East Asia, and Africa, Evans (2009: 1029) could show what also my cases exemplified: interventions on the metropolitan level in order to support creative industries are feasible and often resemble 'traditional' cluster economic policies. The parameters that were identified both in Evans' study (2009: 1029) and my examples range from property/land/zoning issues and fiscal incentives as well as funding/grants/loans to public service provision as well as hard (physical) and soft (such as education) infrastructure. Applying this to Florida's argumentation, the role of transportation infrastructure is only one example where he misperceives the importance of traditional locational factors (2003: 9). Florida also misses that some factors (such as history, weather, environmental endowment etc.) are beyond the scope of most/any actors – Polèse and Shearmur (2006), for instance, describe this aspect by showing the limited scope of economic development policies in declining Canadian regions due to demographic factors.

Another conclusion I would like to draw, concerns Florida's notion of cities needing a "people climate even more than they need a business climate" (2002: 283), Walcott's explanation for the success of San Diego's bioscience cluster due to the specific "regional culture" (2002:

111), or the notion of Berlin as a "cosmopolitan, tolerant and cool city" (Hollersen and Kurbjuweit 2011). I see the "people climate" as highly dependent not on the city governments' politics but the corresponding companies' or clusters' culture. Creative industries are so centered on the individual creative mind that the relation between employee and firm or the lifestyles in a cluster/scene will make up a "regional culture" or a "cosmopolitan, tolerant and cool city". This means that a city government with the scope of its policies can best influence the "business climate", while the "people climate" is coined by the creative industries, because in these industries, creative people make up and form the culture of their sector and maybe the city.

There are, however, questions left: I have shown that creative industries can depend on factors that are influenced on various scales. These industries do also often cross political boundaries, thus acting on various scales. In my introductory remarks, I have said that I am missing the link between creative city research and cluster economics research. As other authors have pointed out, we need to improve our understanding of state and federal-level policies on scale-crossing (creative) industries (Delgado, Porter, and Stern 2011: 33). Evans argues that policies in support of creative industries pretty much resemble 'good old' start-up and SME policies (2009: 1029). And Pratt correctly recognizes that creative city's characteristics pretty much resemble a re-visioning of the agenda for livable cities (Pratt 2008: 6). This leads us to the question what difference the notion of creativity makes with regard to (the scope of) urban policies? Research has looked into the important and promising role of various governance options. But focusing on city governments' policies towards clusters, creative entrepreneurs, and high-quality urban life, it seems to be the next task to bring these related research areas together.

5. References

Bathelt, Harald, Anders Malmberg, and Peter Maskell (2004): Clusters and knowledge: local buzz, global pipelines and the process of knowledge creation. *Progress in Human Geography* 28 (1): 31-56.

Bennett, Darryn (2008): How San Diego Biotech Started and Where It's Going. *Voice of San Diego*: 4 August 2008. Internet: http://www.voiceofsandiego.org/news/article_a9a11f8e-5528-566a-948f-6677ab870b2e.html (11 March 2012).

Benz, Arthur and Nicolai Dose (2010): Governance – Modebegriff oder nützliches sozialwissenschaftliches Konzept?. In: Arthur Benz and Nicolai Dose (Eds.) : *Governance – Regieren in komplexen Regelsystemen*. Second Edition. Wiesbaden (VS): 13-36.

Berridge, Joe (2006): *The Creative City*. Speech at the Mayor's Lunch for Business and the Arts in Regina: 21 March 2006. Internet: http://metrohamilton.ning.com/profiles/blogs/the-creative-city-joe (11 March 2012).

BIOCOM (2006): *Issues and Principles Agenda*. Internet: http://www.biocom.org/documents/BIOCOM_issues_and_prin_06.pdf (11 March 2012).

BIOCOM (2010): *Strategic Plan 2010-2012*: Internet: http://www.biocom.org/?m=sp_view_doc&file=Shared%20Documents/Images/Home%20page/BIOCOM_StrategicPlan2010-2012.pdf (11 March 2012).

Bürkner, Hans-Joachim (2009): Der lokale Staat als Akteur im Feld kreativer Nischenökonomien. In: Bastian Lange, Ares Kalandides, Birgit Stöber, and Inga Wellmann (Eds.): *Governance der Kreativwirtschaft: Diagnosen und Handlungsoptionen*. Bielefeld (transcript): 247-260.

City of San Diego (2008): *City of San Diego General Plan*. Internet: http://www.sandiego.gov/planning/genplan/ (11 March 2012).

City of San Diego Commission for Arts and Culture (2011): *The 2010 Economic and Community Impact of 70 Nonprofit Arts and Culture Organizations*. Internet: http://www.sandiego.gov/arts-culture/pdf/110407report.pdf (11 March 2012).

CONNECT (2012a): *About CONNECT*. Internet: http://www.connect.org/about/ (11 March 2012).

CONNECT (2012b): *Annual Report 2011*. Internet: http://www.connect.org/about/AR2012.pdf (11 March 2012).

Delgado, Mercedes, Michael E. Porter, and Scott Stern (in press): *Clusters, Convergence, and Economic Performance*. Harvard Business School – Institute for Strategy and Competitiveness. Internet: http://www.isc.hbs.edu/pdf/DPS_Clusters_Performance_2011-0311.pdf (11 March 2011).

DeMaio, Carl (2012): Pathway to Prosperity. *Voice of San Diego*: 27 February 2012. Internet: http://www.voiceofsandiego.org/pdf_18e2fc3c-61aa-11e1-a30e-0019bb2963f4.html (11 March 2012).

Dillon, Liam (2012): Mayoral Candidates Want You to Have a Job. *Voice of San Diego*: 1 March 2012. Internet: http://www.voiceofsandiego.org/government/thehall/article_6747189e-63f9-11e1-8202-001871e3ce6c.html (11 March 2012).

Evans, Graeme (2009): Creative Cities, Creative Spaces and Urban Policy. *Urban Studies* 46 (5&6): 1003-1040.

Fletcher, Nathan (2012): Vision for our Economic Future. *Voice of San Diego*: 27 February 2012. Internet: http://www.voiceofsandiego.org/pdf_b21f3d34-61aa-11e1-8036-0019bb2963f4.html (11 March 2012).

Florida, Richard (2002): *The Rise of the Creative Class*. Paperback edition 2004. New York (Basic Books).

Florida, Richard (2003): Cities and the Creative Class. *City & Community* 2 (1): 3-19.

Global CONNECT (2010): *Biotechnology Cluster Project: San Diego Analysis*. Sydney (United States Studies Centre, University of Sydney).

Groove (2012): Club-Ocalypse Now? Berliner Clubbetreiber und Politiker diskutieren beim #a2n_salon. *Groove*: 20 January 2012. Internet: http://www.groove.de/2012/01/20/club-ocalypse-now-berliner-clubbetreiber-und-politiker-diskutieren/ (11 March 2012).

Hollersen, Wiebke and Dirk Kurbjuweit (2011): A Victim of Its Own Success: Berlin Drowns in Tourist Hordes and Rising Rents. Translated by Christopher Sultan. *SPIEGEL International*: 16 September 2011. Internet: http://www.spiegel.de/international/germany/0,1518,786392,00.html (11 March 2012).

House of Research (2011): *Kultur- und Kreativwirtschaftsindex Berlin-Brandenburg 2011: Wirtschaftliche Stimmung und Standortbewertung*. Berlin (SenWTF).

jof/dino (2010): Berlin verlängert liberale Haschisch-Regelung. *Berliner Morgenpost*: 15 May 2010. Internet: http://www.morgenpost.de/berlin/article1308867/Berlin-verlaengert-liberale-Haschisch-Regelung.html (11 March 2012).

Kunzmann, Klaus R. (2009): Kreativwirtschaft und strategische Stadtentwicklung. In: Bastian Lange, Ares Kalandides, Birgit Stöber, and Inga Wellmann (Eds.): *Governance der Kreativwirtschaft: Diagnosen und Handlungsoptionen*. Bielefeld (transcript): 33-46.

Lange, Bastian, Ares Kalandides, Birgit Stöber, and Harald A. Mieg (2008): Berlin's Creative Industries: Governing Creativity? *Industry and Innovation* 15 (5): 531-548.

Lange, Bastian, Ares Kalandides, Birgit Stöber, and Inga Wellmann (2009a): Diagnosen, Handlungsoptionen sowie zehn abschließende Thesen zur Governance der Kreativwirtschaft. In: Bastian Lange, Ares Kalandides, Birgit Stöber, and Inga Wellmann (Eds.): *Governance der Kreativwirtschaft: Diagnosen und Handlungsoptionen*. Bielefeld (transcript): 325-332.

Lange, Bastian, Ares Kalandides, Birgit Stöber, and Inga Wellmann (2009b): Fragmentierte Ordnungen. In: Bastian Lange, Ares Kalandides, Birgit Stöber, and Inga Wellmann (Eds.): *Governance der Kreativwirtschaft: Diagnosen und Handlungsoptionen*. Bielefeld (transcript): 11-32

Malanga, Steven (2004): The Curse of the Creative Class. *City Journal* 14 (1): 36-45.

Markusen, Ann (2006): *Cultural Planning and the Creative City*. Paper presented at the Annual American Collegiate Schools of Planning meetings in Fort Worth, Texas: 12 November 2006. Internet: http://www.hhh.umn.edu/img/assets/6158/271PlanningCulturalSpace.pdf (11 March 2012).

Mösken, Anne Lena (2011): Die Nöte der Partybranche. *Berliner Zeitung*: 5 November 2011. Internet: http://www.berliner-zeitung.de/berlin/szene-die-noete-der-partybranche,10809148,11106398.html (11 March 2012).

Peck, Jamie (2005): Struggling with the Creative Class. *International Journal of Urban and Regional Research* 29 (4): 740-770.

Polèse, Mario and Richard Shearmur (2006): Why some regions will decline: A Canadian case study with thoughts on local development strategies. *Papers in Regional Science* 85 (1): 23-46.

Popblog (2011): Interview zur Eröffnung des Gretchen Club. *Popblog*: 26 September 2011. Internet: http://pobplog.berliner-zeitung.de/2011/09/26/interview-zur-eroffnung-des-gretchen-club/ (11 March 2012).

Pratt, Andy C. (2008): Creative cities: the cultural industries and the creative class. *Geografiska Annaler: Series B – Human Geography* 90 (2): 107-117.

Rapp, Tobias (2009): *Lost and Sound: Berlin, Techno und der Easyjet*. Frankfurt (Suhrkamp).

Scott, Allen J. (2006a): Creative Cities: Conceptual Issues and Policy Questions. *Journal of Urban Affairs* 28 (1): 1-17.

Scott, Allen J. (2006b): Entrepreneurship, Innovation and Industrial Development: Geography and the Creative Field Revisited. *Small Business Economics* 26 (1): 1-24.

Senatsverwaltung für Wirtschaft, Technologie und Forschung (2012): *Die Berliner Landesinitiative Projekt Zukunft: Dokumentation 1997 – 2011*. Berlin (SenWTF).

Senatsverwaltung für Wirtschaft, Technologie und Frauen (2008): *Kulturwirtschaft in Berlin: Entwicklungen und Potenziale*. Berlin (SenWTF).

Senatsverwaltung für Wirtschaft, Technologie und Frauen (2007): *Studie über das wirtschaftliche Potenzial der Club- und Veranstaltungsbranche in Berlin*. Berlin (SenWTF).

Stollowsky, Christoph (2011): Berliner Nachtleben: Die Politik entdeckt die Clubs. *Tagesspiegel*: 19 October 2011. Internet: http://www.tagesspiegel.de/berlin/berliner-nachtleben-die-politik-entdeckt-die-clubs/5223010.html (11 March 2012).

Teipelke, Renard (2011): Berlin: Until the last basement is found. *Blog Place Management & Branding*: 10 May 2011. Internet: http://blog.inpolis.com/2011/05/10/until-the-last-basement-is-found/ (11.03.2012).

Thierstein, Alain, Agnes Förster, and Stefan Lüthi (2009): Kreativwirtschaft und Metropolregionen – Konturen einer systematischen Steuerung. In: Bastian Lange, Ares Kalandides, Birgit Stöber, and Inga Wellmann (Eds.): *Governance der Kreativwirtschaft: Diagnosen und Handlungsoptionen*. Bielefeld (transcript): 62-86.

Voice of San Diego (2012): *The Road to 2012: Mayor's Race*. Internet: http://www.voiceofsandiego.org/mayor-2012/ (11 March 2012)

Walcott, Susan M. (2002): Analyzing an Innovative Environment: San Diego as a Bioscience Beachhead. *Economic Development Quarterly* 16: 99-114.

Walter, Birgit (2011): Musikstandort Berlin: Wir führen Klubs, keine Bordelle. *Berliner Zeitung*: 28 October 2011. Internet: http://www.berliner-zeitung.de/kultur/musikstandort-berlin-wir-fuehren-klubs--keine-bordelle,10809150,11069692.html (11 March 2012).

Zha, Weixin (2011): Der gutgelaunte Protest. *Tageszeitung (taz)*: 17 July 2011. Internet: http://www.taz.de/!74673/ (11 March 2012).